NEXT-GENERATION MILITARY TECHNOLOGY

AI IN THE MILITARY

by George Anthony Kulz

BrightPoint Press

San Diego, CA

BrightPoint Press

an imprint of ReferencePoint Press, Inc.
Printed in the United States

For more information, contact:
BrightPoint Press
PO Box 27779
San Diego, CA 92198
www.BrightPointPress.com

LIBRARY OF CONGRESS CATALOGING-IN-PUBLICATION DATA

Name: Kulz, George Anthony, author.
Title: AI in the military / by George Anthony Kulz.
Description: San Diego, CA: ReferencePoint Press, 2027 | Series: Next-generation military technology | Includes bibliographical references and index. | Audience: Grades 7–9
Identifiers: ISBN 9781678214203 (hardcover) | ISBN 9781678214210 (eBook)
The complete Library of Congress record is available at www.loc.gov.

CONTENTS

AT A GLANCE

- Artificial intelligence (AI) is the ability of a computer to perform tasks that people would normally do. AI finds patterns and learns through experience.

- AI in the military can make things easier for soldiers, military leaders, and support personnel. It also helps keep people and information safe.

- Autonomous vehicles with AI can control themselves. They carry supplies, help soldiers on the battlefield, and spy on enemies.

- Some militaries are using robots with AI to help train soldiers. These robots are also used for different operations on the battlefield.

- AI can gather information about enemies and battles. It figures out risks and opportunities.

- Special AI programs act as wargames and simulations. These special programs provide a safe way to practice fighting. Military personnel can also try new plans and strategies.
- AI helps with behind-the-scenes military tasks. These include cybersecurity, human resources, and support tasks.
- Some experts believe it is important for people to have oversight of military AI systems. AI can make mistakes, and in the military, mistakes can be deadly.

EYES OVER UKRAINE

A small object flies high over the skies of Ukraine. It cannot be seen by enemy Russian troops. The object is a Ukrainian military drone. It has guns on board in case it is attacked. However, that is not its main purpose. It is there simply to collect information.

Suddenly, enemy troops move out from the cover of trees. The drone rushes into action. It follows the troops' movements.

Drones have many different parts. These parts help drones fly.

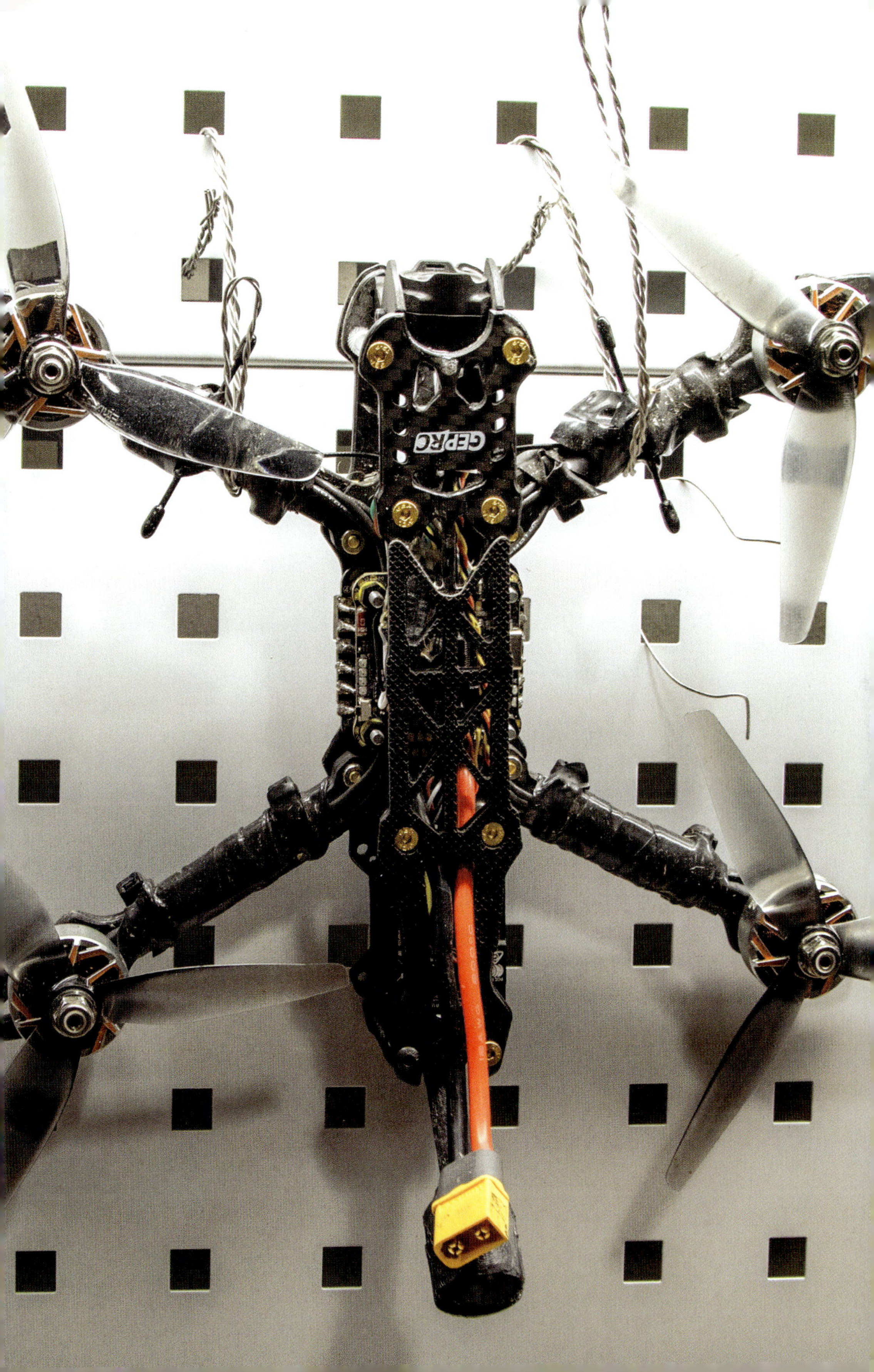
GEPRC

As it flies, an enemy drone zooms over and attacks. The two drones exchange gunfire. The enemy drone scores a direct hit. The drone crashes to the ground. Normally, crashing would be a failure. But this drone has already completed its mission.

The drone was a small helicopter with a computer on board. It was programmed to fly its mission with no pilot. It also managed to send back information on the enemy troops. This includes their location, movements, and weapons. Similar data has been collected by many other drones. Special computer programs process all this information. It is then given to Ukrainian troops. It will aid them in their war against Russia. It will also help with future military efforts.

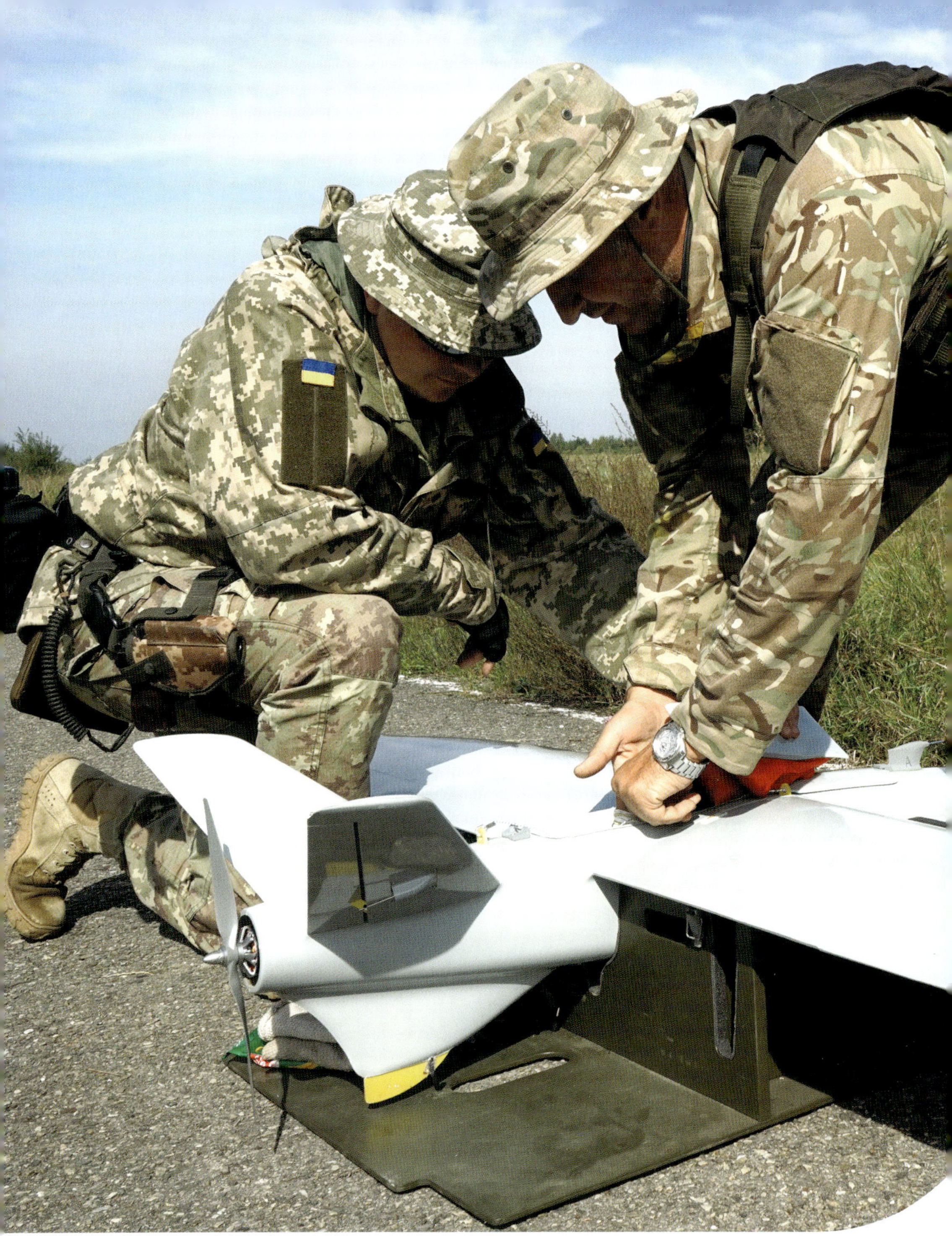

Some drones look like planes. They can be equipped with cameras to observe the battlefield from the sky.

THE USE OF ARTIFICIAL INTELLIGENCE IN THE MILITARY

The drone uses artificial intelligence (AI) to complete its mission. AI is the ability of a computer to perform tasks that humans typically do. It is **software** that runs on

Some robots are controlled remotely. Those powered with AI can move by themselves.

a computer. The human brain is good at making decisions. It sifts through data to find patterns. It also learns from experience. People design AI to do these same things.

Drones are just one example of how AI is used in the military. AI runs on many types of computers. It runs in a **data center** to support cybersecurity. Military AI makes some things faster and more efficient. It also keeps people and information safe.

CHAPTER ONE

MILITARY USE OF AI

In 2024, National Security Advisor Jake Sullivan gave a speech about AI. He talked about the future of AI in the military. He said the military needed new AI tools. If the military did not keep up with AI advances, other nations could use AI against the United States.

Sullivan said the military needed to use AI the right way. AI was growing fast. No one knew exactly what it would become.

National Security Advisor Jake Sullivan delivers a speech at the White House in 2025. During his talk, he discusses the future of technology and AI.

But one thing was clear. Sullivan said, "There is probably no other technology that will be more important to our national security in the years ahead."[1]

AI hardware is put into different technology to help it perform tasks.

THE BENEFITS OF AI

There are many benefits of using AI in the military. It is helpful during battles. AI software runs on machines such as drones and robots. These machines are sent into battle. They search for enemy soldiers and equipment. They send back information on what they see. AI also helps identify what these machines find. Some machines that use AI act as weapons. Others enter risky areas and rescue soldiers. One goal of AI is to keep people safe during combat.

AI is also useful away from combat. It helps with communication. People in charge of the military command those under them. They control everything the military does. They make sure communication is working. AI can help with these functions.

It can coordinate **asset** movements. It can analyze large amounts of data. It suggests ideas and makes predictions.

AI has many noncombat uses. It is used in information warfare. This involves protecting information and spreading false information to mislead the enemy. AI is good for **logistics** support. It can track supply chains. It predicts when and what to order and where to send it. This information helps the military run smoothly.

USES OF AI IN THE MILITARY

AI controls vehicles and robots. Unmanned ground vehicles (UGVs) are land vehicles that move without a driver. They help soldiers on the ground. Unmanned aerial vehicles (UAVs) help from the air.

AI technology can let people know when equipment needs to be fixed or replaced.

Experts are experimenting with robot dogs that can move across different terrain.

An example of a UAV is a drone. Unmanned underwater vehicles (UUVs) work in water. They do jobs that are too dangerous for people.

Military leaders can use AI in many ways. AI programs can process large amounts of data at once. They can help people understand a complicated battlefield. AI finds patterns in the data. It uses patterns to help troops decide what to do next. AI is also used for wargaming and simulation. This helps train leaders and soldiers for real battles.

The military uses AI to help with cybersecurity. Cybersecurity is the process of protecting information. It keeps computers and information safe from hackers.

Military human resources (HR) personnel can also use AI to help complete tasks. AI programs assist with recruiting new troops. They also help with **onboarding** and training, too.

CHALLENGES OF AI

Using AI in the military comes with challenges. Sometimes AI does not have all

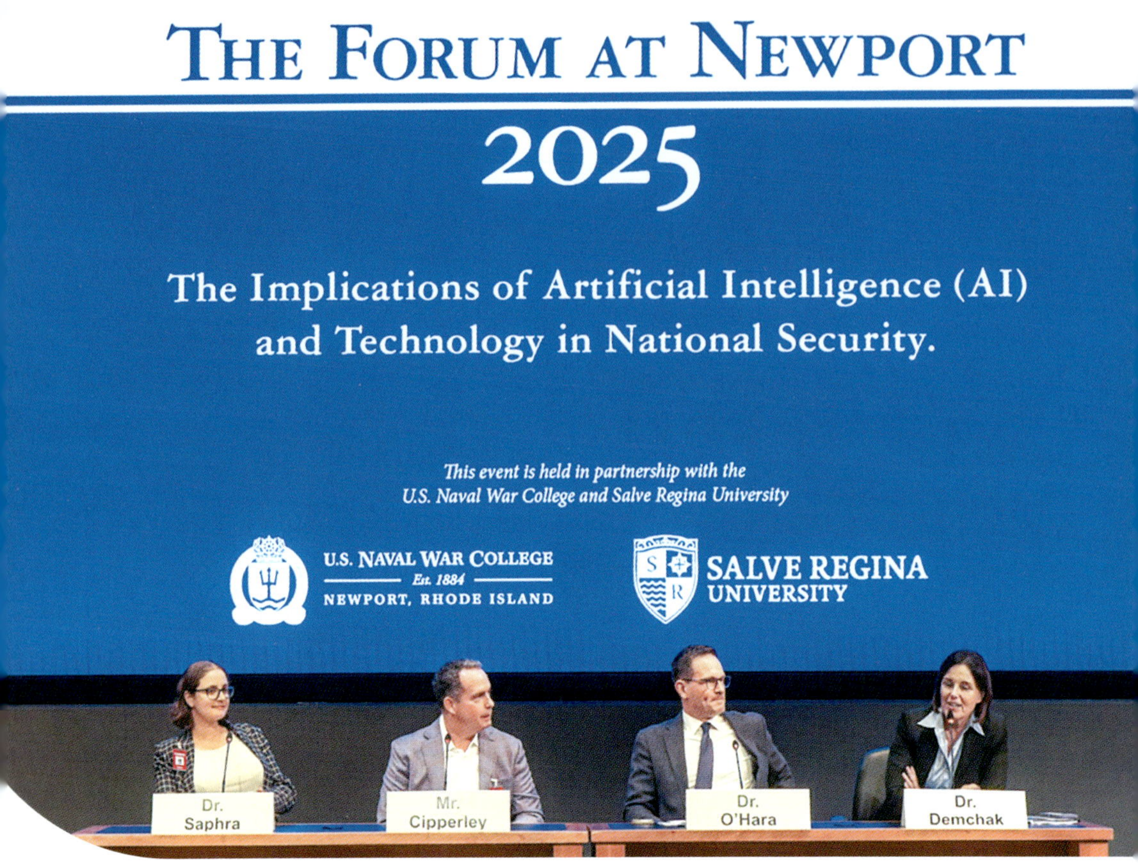

Experts and military personnel talk about the usage of AI and technology in Newport, Rhode Island.

the information it needs to work accurately. AI can also have bad information. This could result in bias or unfair judgment. AI can make choices using only what it has learned. This means it can make mistakes or be unfair.

Mistakes can have deadly results in the military. Choices made by AI can hurt people. AI may make these decisions on its own. It can be hard to know who is responsible if things go wrong. It is important to think carefully about these challenges. By doing so, AI in the military can be used responsibly.

HITL vs. HOTL

There are two main views about how AI fits into the military. The first is humans-in-the-loop, or HITL. In this view, humans should have control over every step AI takes. The second is humans-on-the-loop, or HOTL. In this view, humans oversee AI. They take action only when something goes wrong.

CHAPTER TWO

AUTONOMOUS VEHICLES AND WEAPONS

Many countries use self-driving vehicles in battle. These are known as autonomous vehicles. They use computers and AI to work without human help.

On August 28, 2023, US military official Kathleen Hicks spoke about autonomous weapons. She said, "To stay ahead, we're going to create . . . autonomous systems . . . which are less expensive, put fewer people in the line of fire, and can

Weapons are being developed to have AI technology.

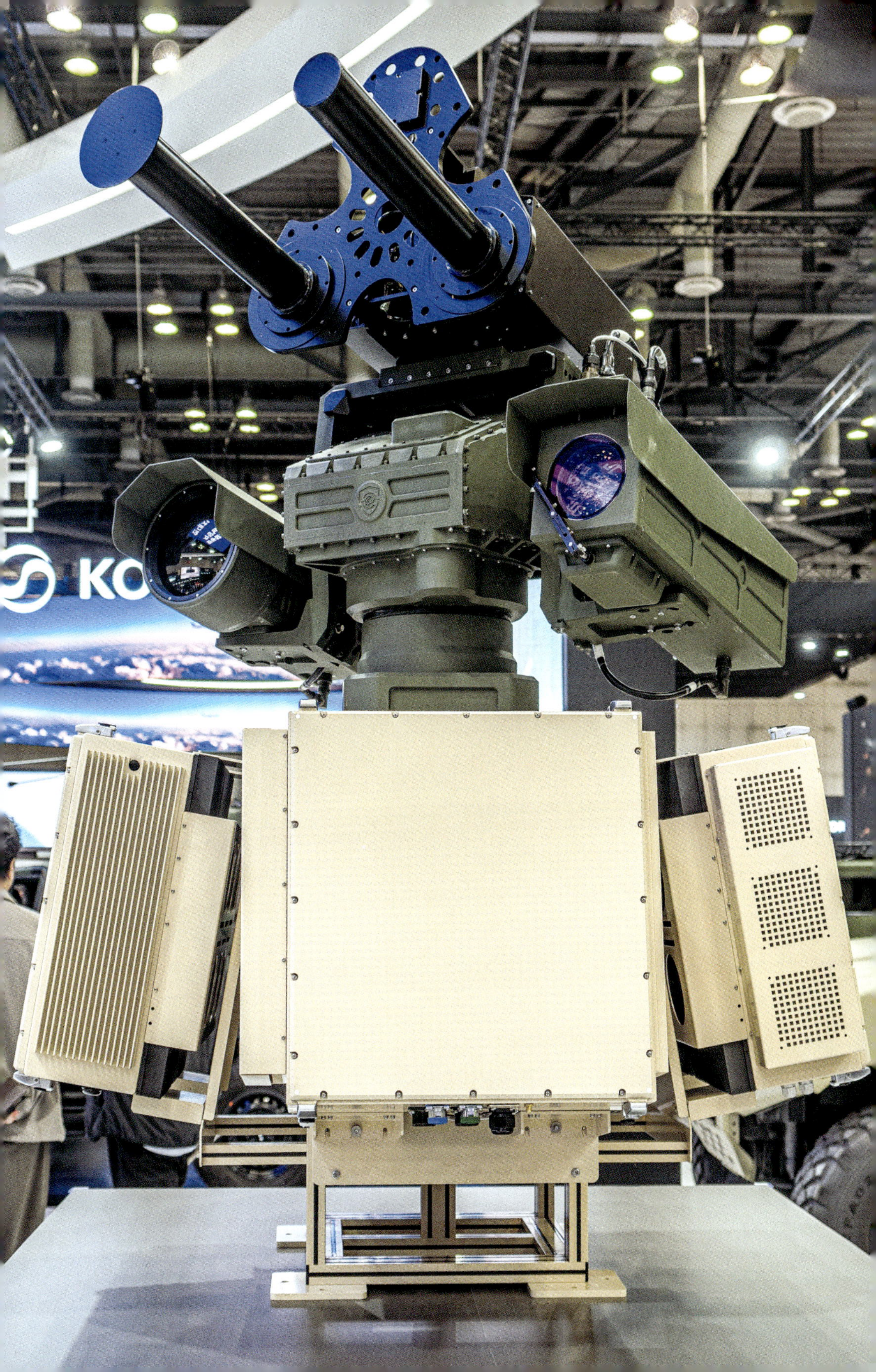

be changed, updated, or improved with substantially shorter lead times."[2]

UNMANNED GROUND VEHICLES (UGVS)

UGVs are autonomous vehicles that move along the ground. Their movement is controlled by AI. This means they can move

Large UGVs can carry supplies and even people.

and make decisions on their own. They may be cars, trucks, or all-terrain vehicles (ATVs).

UGVs use special equipment powered with AI that helps them function by themselves. Some UGVs use Light Detection and Ranging (LiDAR) sensors. These sensors use lasers to see the surroundings. They build a 3D model of the area. AI helps process this information. It then decides which way to go. AI spots important objects and threats. It sends back data for people to analyze. It fights threats by using targeting and guidance systems.

UGVs with AI have some advantages over vehicles controlled by people. They can draw enemy fire away from soldiers. They can find and rescue people. They can do dangerous tasks, such as

Sensors, cameras, and AI programming work together to create accurate images of a battlefield.

handling bombs. UGVs with AI do not get tired like people do. They do not need to take rests. Some small UGVs stay hidden and collect information.

UNMANNED AERIAL VEHICLES (UAVS)

UAVs are autonomous vehicles that fly. Drones are the most common UAVs. Some look like tiny helicopters. Some look like small planes. Their size and shape depend on what they do.

Like UGVs, UAVs may use LiDAR to help sense their surroundings. Sensors help

Some drones can carry supplies over large distances.

them identify objects. They can also communicate what they see to people. They can identify and fight targets. However, there are some differences. UAVs have a longer range than ground vehicles. They reach areas UGVs cannot. They can see a large area all at once.

UAVs with AI have some benefits over aircraft flown by people. They can be smaller and lighter than aircraft with pilots. This helps them stay in the air longer. They cost less to make and use. Because of this, they can go into dangerous places without as much risk. This helps keep people safe. UAVs may work alongside piloted aircraft. This can help increase safety for the human pilots.

UNMANNED UNDERWATER VEHICLES (UUVS)

UUVs are autonomous vehicles that move in water. Their shapes and sizes depend on their use. Some are designed with **biomimicry**. These UUVs look or behave like underwater animals. The Manta Ray

The GhostSwimmer vehicle looks like a shark. It blends into its surroundings to stay hidden.

UUV looks like a manta ray fish. It uses AI to easily navigate through the water.

AI is especially important for UUVs. UUVs do not have reliable communication systems underwater. People cannot see the collected data right away. UUVs must do their own analysis of the data. They have several types of sensors. UUVs use optical, sonar, and acoustic sensors. AI processes the data from these sensors to help UUVs see underwater. It helps them move through the water safely. Some UUVs can learn to navigate underwater better over time.

UUVs are built to withstand lots of pressure. They operate at depths where people cannot go. Some UUVs swim in groups to cover more area and gather more data.

HUMANOID ROBOTS

The future of war might involve humanoid robots with AI. These are robots that look like humans. As of 2025, no humanoid robots have appeared on the battlefield. However, work in this area is being done.

In the United States, humanoid robots are being used to help soldiers train for real-world combat scenarios. One such robot made for the US military is

Boston Dynamics

United States company Boston Dynamics is a leader in creating robots with AI. However, the military cannot use them as weapons. The company has a policy against adding weapons to its robots. It says the risks are too great.

the Phantom MK1. It was made by the Foundation Robotics Lab.

Robots with AI can have many advantages over human soldiers. They act

United States company Boston Dynamics released a robot called Atlas. It uses AI to learn new skills.

as targets for soldiers during training. They can also act as distractions in battles to draw fire away from humans. They can move through areas that are hard for people. The long-term cost of using AI robots instead of humans is much less. The biggest benefit is fewer people get hurt.

Using robots with AI comes with many **ethical** concerns. Robots might make biased decisions. They might hurt the wrong targets. Naveen Joshi is a team lead at Allerin. It is a software company. He says, “We will always have two factions, one for and one against the use of robots in wars. However, the use of robots in combat is inevitable.”[3]

COMMAND, CONTROL, AND DECISION SUPPORT

Command, control, and decision support are all important tasks in the military. *Command* means leaders tell troops what to do. *Control* means leaders use plans and tools to guide missions. *Decision support* means leaders get help from tools so they can make smart choices. Computers with AI can show what is happening. AI can offer plans and help people find the best option.

Soldiers will need to be trained on how to use AI technology.

Getac

In 2020, US Air Force General John E. Hyten spoke about the battlefields of the future. He said,

> *If we go into a future where there are no lines on the battlefield, and we have* ***ubiquitous*** *all-domain command and control and logistics that go seamlessly from place to place . . . and it all happens in enormous speed . . . that is the world where an* ***adversary*** *will not challenge us. . . . I can't figure out in my mind how to do that without artificial intelligence.*[4]

SITUATIONAL AWARENESS

AI is helping with situational awareness. Situational awareness means knowing

AI can help leaders make tough decisions.

what is happening nearby. In the military, situational awareness has three parts. First, leaders gather information. Second, they study that information. They notice things that are important. Finally, they use that information to figure out the overall situation. These three things are needed to form a good plan for what to do next.

All of these things were done only by people. Soldiers had to go into dangerous areas and observe. They analyzed what

Navigating harsh terrain such as jungles can be difficult. AI-powered robots can help people gather information in these conditions.

they saw. Then they reported back to their leaders. Their leaders used this information to form a complete picture. Time and experience were needed to do this effectively.

AI helps make these tasks easier and faster. Autonomous vehicles can gather data. This means people do not have to go to dangerous places. Robotic vehicles can go places people cannot. AI can quickly understand what these vehicles see. Then, AI software combines data from different sources. This gives leaders a clear and complete picture of what is happening. Leaders then decide what to do next.

COURSE OF ACTION

A course of action is a step-by-step mission plan. Military personnel come up with different ideas. They study whether each step will work. They then choose the best option.

Before AI, people decided on courses of action by themselves. They used their own experience and knowledge. Sometimes they relied on their own thinking when faced with new situations. Leaders also asked experts for advice. They learned from what happened in past missions to make better plans.

AI is starting to help with many of these processes. It gathers past and current information from datasets. It identifies risks and opportunities. Finally, it gives leaders different plans to choose from. But people still make the final choice.

Joint All-Domain Command and Control (JADC2)

Joint All-Domain Command and Control (JADC2) is a plan for the US military to use AI to connect all its tools, such as sensors and shooters. JADC2 will gather data from land, sea, space, and cyberspace. It will help leaders see what is going on, understand it, and then act to protect people.

WARGAMING AND SIMULATION

Sometimes military leaders want to test their plans before a real battle. They can use wargames to help. These are pretend battles based on known information about friendly and enemy forces. Military leaders play the roles of both sides. They make decisions about where to send troops, ships, and planes. Wargames help leaders test ideas and practice making choices. They are safer and cost less than real fighting.

AI wargames are different from regular wargames. Regular wargames use set scenarios. They also rely on teachers who lead the games. AI wargames are more flexible. The computer changes the game as people make choices. It copies the way enemies think. AI can even talk with players in different languages. The AI program learns like people do. It does so through trial and error.

The AlphaDogfight Trials were an example of a wargame involving AI. A dogfight is a fight between fighter pilots. The wargame used virtual reality. Some of the players were human pilots. Others were AI programs made by different companies. Colonel Dan "Animal" Javorsek was a program manager in the test. He said, "The goal was to earn the respect of a fighter pilot . . . by demonstrating that an AI agent can quickly and effectively learn basic fighter maneuvers and successfully employ them in a simulated dogfight."[5] The use of AI and special programs is still growing.

TAKING A COMPLETE PICTURE WITH AI

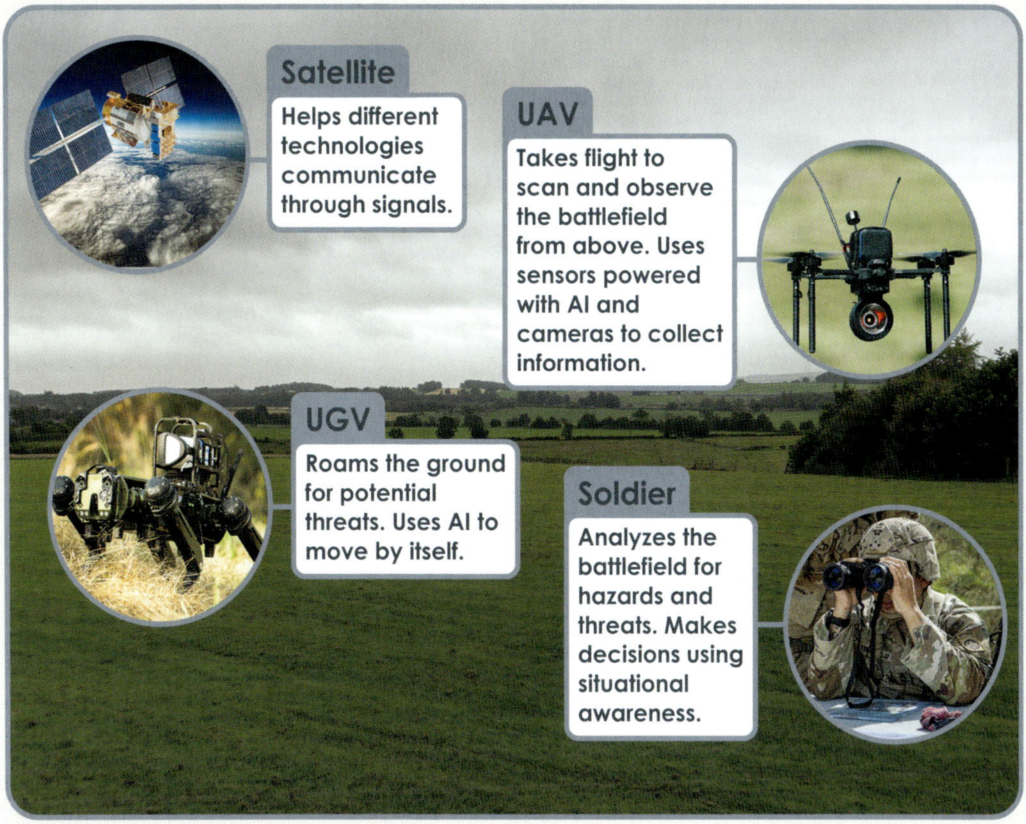

This infographic shows different ways AI and people collect information to create a complete picture of a battlefield.

Humans will need to get used to working with AI on and off of the battlefield. Trainings will help.

NONCOMBAT APPLICATIONS

AI helps the military in many ways beyond the battlefield. It protects computers and keeps information safe. It helps train soldiers. Human resources and other support people also use AI. They may use it to help soldiers and veterans. AI is used to keep track of supplies, too. It makes sure supplies get to where they need to go. It also checks when equipment needs repair.

Virtual reality headsets use AI to build simulations. These simulations can help soldiers train for different tasks.

LCSRON ONE
U.S.
MIRELES

When people use AI, they must do it in a safe and ethical way. Ruksana Lodi works with logistics in the US military. She says, "The goal is to ensure that

The 175th Cyberspace Operations Group Capture the Flag event was held in Maryland. It helped military teams improve their cybersecurity skills.

AI-driven decisions don't compromise security, quality, or operational efficiency, ultimately protecting both the agency and the warfighter."[6]

CYBERSECURITY AND INFORMATION MANAGEMENT

Not all battles are fought on a physical battlefield. Some are fought in the online world of computer networks. This is called the cyberspace. Cybersecurity involves protecting the cyberspace. These include a country's computers, online networks, and information. Some battles also involve attacking the computers and networks of other countries. This is known as information warfare. Cybersecurity also protects against hackers. Hackers break into computer systems. They often steal data or damage the systems by leaving viruses. A virus is a threat that can damage computers.

Before using AI, cybersecurity was done by putting special programs on computers. Networks were set up to protect against attacks. People had to make updates from time to time. They would also have to check computers for suspicious behavior. Today, AI software updates systems. AI spots serious threats faster than people. Paul M. Nakasone is a retired US Army general. He says: "We are experimenting and developing 'self-healing networks,' where we see a vulnerability and the vulnerability is recognized rapidly and patched or mitigated."[7]

Passwords help protect important information. Using unique passwords helps make passwords stronger.

In the past, people did all the work to spread disinformation. They used leaflets, TV broadcasts, and social media. They captured audio, video, or computer data to get military secrets. Today, information warfare is more focused. AI programs learn things about people from online networks. They can send people false information to trick them. AI can set up fake websites and send emails to fool people into sharing sensitive data. AI can even create fake audio, video, and images of people.

HR, SUPPORT, AND TRAINING

Military HR is in charge of finding new people for military positions. HR personnel are responsible for training soldiers and other professionals. They keep track of military personnel. Other support groups take care of wounded soldiers and veterans who need help.

Recruiters are people responsible for finding new soldiers. They must find soldiers who meet certain requirements. These future soldiers must also complete special training. This helps new soldiers learn about the army. This is usually done with

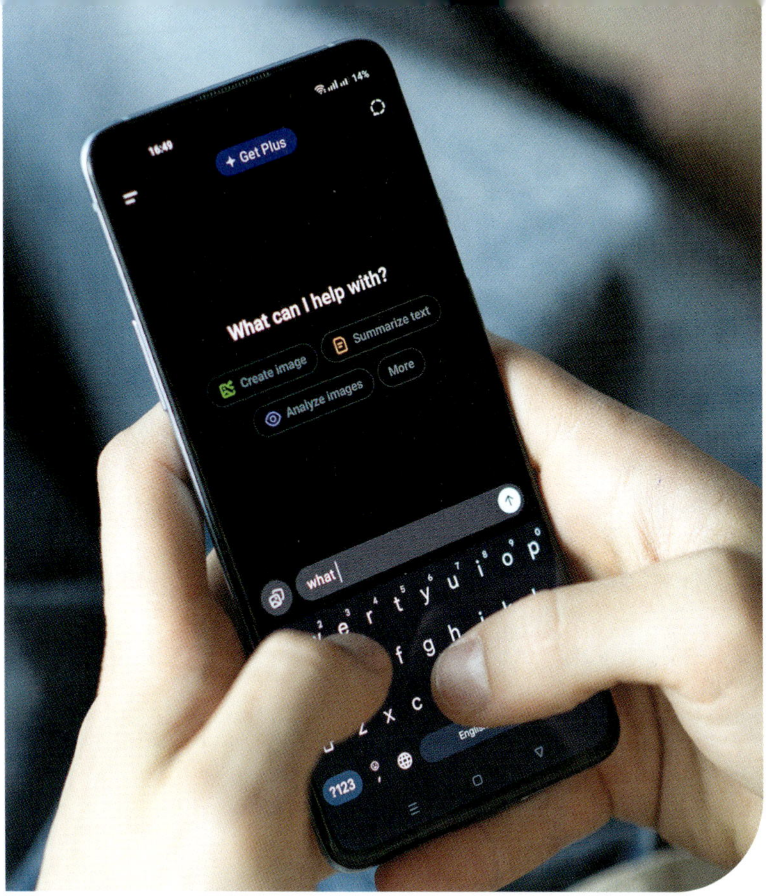

Chatbots are computer programs that can have text-based conversations. AI chatbots can answer questions for soldiers and veterans.

experienced teachers. But sometimes there are not enough workers to do these jobs.

AI has helped solve some of these problems. AI finds ideal people from job applications. It searches for keywords and skills. It schedules interviews online. AI also helps with training. It uses realistic simulations. It quickly scores lessons and gives feedback. AI can also check soldiers' health. Wearable AI devices check vital signs. AI quickly scans images and health data. It finds issues that need attention.

LOGISTICS AND SUPPLY CHAIN

Some important processes in the military involve logistics and the supply chain. Keeping track of military assets is important. It makes sure soldiers are safe and successful at work. Accurate inventory is important. Predicting what soldiers need is also important. Determining when to send supplies to troops is key to successful missions. All equipment should

Some barcode scanners use AI to help keep track of inventory. AI can scan damaged barcodes and fill in the missing information.

be checked regularly. This ensures it works when needed.

People are responsible for these tasks. Inventory is based on current needs. Knowing where supplies are located can be hard. Getting supplies to where they

Condition-Based Maintenance Plus (CBM+)

The Condition-Based Maintenance Plus (CBM+) plan was started by the US Air Force. It is used to keep military equipment working. AI is a key part of this plan. AI uses sensor data from equipment. It looks at the information to predict when repairs or fixes will need to occur. Motor vehicles need oil replacements. They have set schedules for when they need to be repaired. But the CBM+ plan will help replace the oil only when a replacement is needed.

are needed can mean long and dangerous routes. Needed supplies may not be available right away. And equipment gets fixed only when it breaks.

AI helps with logistics and supplies. AI tracks inventory, predicts needs, and finds safe delivery paths. AI ensures equipment check-ups are scheduled. Smart sensors or robots with AI check equipment. They check for damage and other issues. All of this makes sure the people using supplies are safe.

THE FUTURE OF AI

AI is a powerful tool. It can support different military missions. It can also help with daily military processes. However, it should be used with caution. AI lacks the empathy, training, and experience of people. It cannot always provide the best decision for large or complex situations.

AI technology can make tasks easier. At Hill Air Force Base, a machine dispenses medicine quickly and efficiently using AI technology.

In 2025, Joseph L. Huitt was a US lieutenant colonel. He talked about AI in decision-making. Huitt said,

> *To continue our military prowess, artificial intelligence should never replace the critical human element in leadership decision-making. There must always be a human-in-the-loop.*[8]

GLOSSARY

adversary

an enemy or opponent

asset

something owned or controlled by the military

biomimicry

designing something in a way that copies features from animals or plants

data center

a building where computer systems and other data-storing technologies are located

ethical

accepted rules of behavior

logistics

coordination of multiple parts of an organization

onboarding

the process of getting a new employee familiar with their job

software

programs for a computer

ubiquitous

existing or being everywhere

SOURCE NOTES

CHAPTER ONE: MILITARY USE OF AI

1. Quoted in Gerhard Peters and John T. Woolley, "Remarks by National Security Advisor Jake Sullivan on Artificial Intelligence and National Security at the National Defense University," *The American Presidency Project*, October 24, 2024. www.presidency.ucsb.edu.

CHAPTER TWO: AUTONOMOUS VEHICLES AND WEAPONS

2. Quoted in Thomas X. Hammes, "Autonomous Weapons are the Moral Choice," *Atlantic Council*, November 2, 2023. www.atlanticcouncil.org.

3. Quoted in Naveen Joshi, "Is It Ethical to Use Robots in War? What Are the Risks Associated with It?," *Forbes*, July 25, 2022. www.forbes.com.

CHAPTER THREE: COMMAND, CONTROL, AND DECISION SUPPORT

4. Quoted in John E. Hyten, "Remarks by General John E. Hyten to the Joint Artificial Intelligence Symposium and Exposition," *US Department of War*, September 9, 2020. www.war.gov.

5. Quoted in "AlphaDogfight Trials Foreshadow Future of Human-Machine Symbiosis," *DARPA*, August 26, 2020. www.darpa.mil.

CHAPTER FOUR: NONCOMBAT APPLICATIONS

6. Quoted in Beth Reece, "AI to Boost Efficiency, Optimize Logistics Support as DLA Standardizes Use of New Tech," *Defense Logistics Agency*, March 17, 2025. www.dla.mil.

7. Quoted in Paul M. Nakasone, "An Interview with Paul M. Nakasone," *Joint Force Quarterly*, no. 92, 2019 p. 4–9. https://ndupress.ndu.edu.

8. Quoted in Joseph L. Huitt, "Leadership: Artificial Intelligence in Decision-Making," *Line of Departure*, July 1, 2025. www.lineofdeparture.army.mil.

FOR FURTHER RESEARCH

BOOKS

Cynthia Kennedy Henzel, *Cyber Warfare in the Military.* BrightPoint Press, 2027.

George Anthony Kulz, *Robotics in the Military.* BrightPoint Press, 2027.

Emily Schlesinger, *Artificial Intelligence.* Saddleback Educational Publishing, 2024.

INTERNET SOURCES

"15 Most Incredible Uses of AI in Military Technology," *Science News Today*, October 9, 2025. www.sciencenewstoday.org.

Debora Henley & 505th Command and Control Wing, "Air Force Experiments With AI, Boosts Battle Management Speed, Accuracy," *US Department of War*, September 23, 2025. www.war.gov.

Charles Horn, "Sharper: Military Artificial Intelligence," *CNAS*, March 13, 2025. www.cnas.org.

WEBSITES

AI Youth
https://aiyouth.org

AI Youth provides free resources for students learning about AI. The website also includes information about educational courses and competitions.

DoD STEM
https://dodstem.us

DoD STEM provides resources related to STEM topics, including robotics, cybersecurity, and AI. The website also provides information on how these topics apply to the military.

Science Journal for Kids and Teens
www.sciencejournalforkids.org

Science Journal for Kids and Teens contains hundreds of science articles for kids and teens, including articles about AI.

INDEX

IMAGE CREDITS

Cover: © TSViPhoto/Shutterstock Images
5: © Gorodenkoff/Shutterstock Images
7: Maj. Joe Legros/US Army/DVIDS
9: Army Spc. Amy Carle/US Army National Guard/DVIDS
10: Sgt. 1st Class Shane Smith/US Army National Guard/DVIDS
13: © Andrew Harnik/Getty Images News/Getty Images
14: Spc. Wyatt Moore/28th Public Affairs Detachment/US Army/DVIDS
17: Airman 1st Class Chelsea Arana/US Air Force/DVIDS
18: Airman 1st Class Tiffany Price/US Air Force/DVIDS
20: Kristopher Burris/US Navy/DVIDS
23: © Flying Camera/Shutterstock Images
24: Sgt. Alfonso Livrieri/US Marine Corps/DVIDS
26: Sgt. Stephen J. Otero/US Air Force/DVIDS
27: Mass Communication Specialist 3rd Class Edward Guttierrez III/US Navy/DVIDS
29: Northrop Grumman/DVIDS
32: © Tayfun Coskun/Anadolu/Getty Images
35: Pfc. Natalia Hernandez/US Army/DVIDS
37: Barry Bena/US Air National Guard/DVIDS
38: Staff Sgt. Andre Taylor/US Army/DVIDS
43 (satellite): © Andrei Armiagov/Shutterstock Images
43 (UAV): US Marine Corps/DVIDS
43 (UGV): Senior Airman William Pugh/US Air Force/DVIDS
43 (soldier): Pfc. Seu Chan/US Army/DVIDS
43 (background): © Rob Atherton/Shutterstock Images
45: Mass Communication Specialist 1st Class Josh Coté/US Navy/DVIDS
46: Staff Sgt. Laura Virtue/US Air National Guard/DVIDS
49: © Patcharin.In/Shutterstock Images
51: © Yarrrrrbright/Shutterstock Images
53: Senior Airman Rachel Coates/US Air Force/DVIDS
56: Cynthia Griggs/US Air Force/DVIDS

ABOUT THE AUTHOR

George Anthony Kulz is a member of the Society of Children's Book Writers and Illustrators and has taken courses at the Institute of Children's Literature and the Gotham Writers' Workshop. He writes for children and adults.